The Baculum in the of Western North America

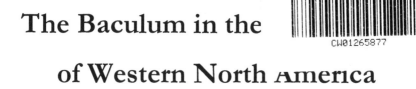

John A. White

Alpha Editions

This edition published in 2024

ISBN : 9789366388052

Design and Setting By
Alpha Editions
www.alphaedis.com
Email - info@alphaedis.com

As per information held with us this book is in Public Domain.
This book is a reproduction of an important historical work. Alpha Editions uses the best technology to reproduce historical work in the same manner it was first published to preserve its original nature. Any marks or number seen are left intentionally to preserve its true form.

INTRODUCTION

The baculum is the bony part of the penis. In the species of the subgenus *Neotamias* the proximal part of the baculum is termed the shaft, and the distal upturned part is termed the tip. On the dorsal side of the tip there is a longitudinal ridge termed the keel. The proximal end of the shaft is termed the base (see fig. 19). Depending on the species, the shaft varies from 2.11 to 5.28 mm. in length, and the base may or may not be widened or deepened.

The purpose of this report is to: (1) Show the usefulness of the structure of the baculum as a taxonomic character in chipmunks; and (2) compare a classification based on the structure of the baculum with a classification based on the structure and appearance of the skull and skin.

METHODS, MATERIALS, AND ACKNOWLEDGMENTS

The bacula which were borrowed from the University of Michigan, Museum of Zoology, were processed according to the method described by Friley (1947:395-397), whereas all others were processed according to the method described by White (1951:125). Thus the bacula that were borrowed from the University of Michigan, are maintained there in a separate collection, whereas the bacula borrowed from other museums and those that are at the University of Kansas, Museum of Natural History, are housed with the skulls of the corresponding specimens.

All measurements of the bacula were made by means of an eyepiece micrometer.

A total of 194 bacula were seen. All of these are in the Museum of Natural History of the University of Kansas, unless otherwise indicated by the following symbols:

BS   United States Biological Surveys Collection.

CN   Chicago Natural History Museum.

LA   Los Angeles County Museum.

MM University of Michigan, Museum of Zoology.

NM United States National Museum.

UU   University of Utah, Museum of Zoology.

I am grateful to Professor E. Raymond Hall for guidance in my study and thank Drs. Robert W. Wilson, Keith R. Kelson, and Edwin C. Galbraith, as well as other friends and associates of the Museum of Natural History, University of Kansas, for encouragement and valuable suggestions. Dr. William L. Jellison, United States Public Health Service, aided me in part of my field work and kindly sent me some specimens of chipmunks. My wife, Alice M. White, made the illustrations and helped me in many ways.

For the loan of bacula I thank Dr. William H. Burt, University of Michigan, Museum of Zoology. For permission to search for bacula on study skins, and to process those that were found, I thank Miss Viola S. Schantz, United States Fish and Wildlife Service,

Mr. Colin C. Sanborn, Chicago Natural History Museum, Mr. Kenneth E. Stager, Los Angeles County Museum, Dr. David H. Johnson, United States National Museum, and Dr. Stephen D. Durrant, Museum of Zoology, University of Utah.

Assistance with field work is acknowledged from the Kansas University Endowment Association, the National Science Foundation and the United States Navy, Office of Naval Research, through contract No. NR 161 791.

VARIATION

Individual variation.—Individual variation is small. This is shown by a coefficient of variability of only 3.85 in the length of the shaft in a series of 12 specimens of *E. umbrinus umbrinus* from Paradise Park, 21 mi. W and 15 mi. N Vernal, 10,050 ft., Uintah County, Utah.

Variation with age.—In the chipmunks the baculum varies but little with age. In the youngest specimens that I have taken, the M3 and m3 have not yet erupted and there is no wear on P4 and p4; nevertheless, the baculum in these specimens has nearly an adult configuration and size. In juvenal *Eutamias minimus* the tip of the baculum is longer in relation to the length of the shaft than it is in adults; the tip is 18 to 28 per cent of the length of the shaft in adults, as opposed to 29 to 32 per cent in juveniles.

Aberrations.—In a small percentage of specimens of *E. minimus* and *E. umbrinus* the baculum is small and S-shaped, even in adults.

Variations of taxonomic worth.—Variations in this category are described in the section immediately following the key.

KEY TO THE BACULA IN EUTAMIAS OF WESTERN NORTH AMERICA

- 1. Distal 1/2 to 2/3 of shaft markedly compressed laterally; base markedly widened.

-
  - 2. Distal 1/2 of shaft laterally compressed and curved downward to base of tip.
  -
    - 3. Height of keel 1/2 of length of tip; keel markedly enlarged.
      - *Eutamias bulleri*, p. 627
    - 3'. Height of keel 1/4 of length of tip; keel not markedly enlarged.
      - *Eutamias umbrinus* and *E. palmeri*, pp. 626, 627
  - 2'. Distal 2/3 of shaft laterally compressed and curved downward to base of tip.
  -

- 4. Base of keel 1/3 of length of tip; angle formed by tip and shaft less than 100°
  - *Eutamias speciosus*, p. 625
- 4'. Base of keel 1/2 of length of tip; angle formed by tip and shaft more than 102°
  - *Eutamias panamintinus*, p. 625

- 1'. Distal 1/12 to 2/5 of shaft slightly compressed laterally; base not markedly widened.

  - 5. Shaft thin; shaft less than .20 mm. in diameter at widest point.

    - 6. Ridges on either side of keel enlarged, partially obscuring lateral view of keel; height of keel 1/10 of length of tip.
      - *Eutamias sonomae*, p. 619
    - 6'. Ridges on either side of tip not enlarged, not partially obscuring lateral view of keel; height of keel at least 1/7 of length of tip.

    - 7. Base not widened or dorsoventrally thickened.

    - 8. Shaft more than 4.5 mm. in length; tip 16 per cent of length of shaft; shaft strongly incised on dorsal side of base
      - *Eutamias merriami*, p. 621
    - 8'. Shaft less than 4.4 mm. in length; tip more than 25 per cent of length of shaft; shaft not incised on dorsal side of base.

    - 9. Height of keel 1/7 of length of tip; angle formed by tip and shaft distinct
      - *Eutamias alpinus*, p. 616

9'. Height of keel at least 1/5 of length of tip; angle formed by tip and shaft poorly defined.

10. Height of keel 1/3 of length of tip; angle formed by tip and shaft 140°

Eutamias dorsalis, p. 620

10'. Height of keel 1/5 of length of tip; angle formed by tip and shaft 130° or less.

11. Tip more than 29 per cent of length of shaft.

Eutamias amoenus, p. 619

11'. Tip less than 28 per cent of length of shaft.

Eutamias minimus, p. 617

7'. Base widened and dorsoventrally thickened.

Eutamias townsendii, p. 618

5'. Shaft thick; shaft more than .25 mm. in diameter at widest point.

12. Length of shaft less than 3.00 mm.; length of tip less than 1.10 mm.

Eutamias quadrivittatus hopiensis, p. 622

12'. Length of shaft more than 3.10 mm.; length of tip more than 1.15 mm.

13. Tip less than 28 per cent of length of shaft.

Eutamias quadrimaculatus, p. 624

13'. Tip more than 29 per cent of length of shaft.

14. Angle formed by tip and shaft more than 140°; ridges on either side of tip indistinct.

Eutamias cinereicollis, p. 624

14'. Angle formed by tip and shaft less than 135°; ridges on either side of tip distinct.

15. Shaft less than 3.65 mm. in length, and .55 mm. or less in diameter at widest point.

Eutamias quadrivittatus quadrivittatus, p. 621

15'. Shaft usually more than 3.65 mm. in length, but when shorter, diameter is .60 mm. or more at widest point.

16. Diameter of shaft at widest point less than .58 mm.; tip less than 35 per cent of length of shaft.

Eutamias ruficaudus ruficaudus, p. 622

16'. Diameter of shaft at widest point more than .65 mm.; tip more than 40 per cent of length of shaft.

Eutamias ruficaudus simulans, p. 623

ACCOUNTS BY SPECIES

Eutamias alpinus (Merriam)

Figure 1

Pelage silky; tail bright orange beneath; markings relatively obscure; size small; skull broad, flattened, and large in proportion to body.

Baculum: Shaft thin; keel low, 1/7 of length of tip; tip 39 per cent of length of shaft; angle formed by tip and shaft 135°; distal 1/3 of shaft slightly compressed laterally; base slightly wider than shaft; shaft short, 2.17 mm.

Differs from *E. speciosus*, *E. panamintinus*, *E. umbrinus*, *E. palmeri*, and *E. bulleri*, in base not markedly widened, and shaft thinner; from *E. quadrivittatus*, *E. ruficaudus*, *E. cinereicollis*, and *E. quadrimaculatus*, in shaft thinner, baculum shorter; from *E. townsendii*, in base not dorsoventrally thickened, base not so widened; from *E. sonomae*, in ridges on either side of tip not enlarged, base not dorsoventrally thickened; from *E. amoenus*, *E. minimus*, and *E. dorsalis*, in keel lower, angle formed by tip and shaft more distinct; from *E. merriami*, in baculum markedly shorter, base not incised dorsally.

Specimen examined: One from Big Cottonwood Meadows, S of Mount Whitney, 10,000 ft., Inyo Co., California (CN).

Eutamias minimus (Bachman)

Figure 2

Coloration varying from light to dark depending on subspecies; size small to medium; rostrum short and stout.

Baculum: Shaft thin; keel low, 1/5 of length of tip; tip 18 to 28 per cent of length of shaft; angle formed by tip and shaft 125°; distal 1/2 of shaft slightly compressed laterally; shaft short to long, 2.44 to 4.35 mm.

Differs from *E. speciosus*, *E. panamintinus*, *E. umbrinus*, *E. palmeri*, and *E. bulleri*, in shaft thinner, base not markedly widened; from *E. quadrivittatus*, *E. ruficaudus*, *E. cinereicollis*, and *E. quadrimaculatus*, in shaft thinner, tip shorter; from *E. amoenus*, in tip less than 28 per cent of length of shaft; from *E. dorsalis*, in angle formed by tip and shaft smaller; from *E. townsendii*, in tip less than 28 per cent of length of shaft, angle formed by tip and shaft 125° instead of 130°; from *E. sonomae*, in ridges on either side of tip less well-developed, keel higher; from *E. merriami*, in shaft shorter (less than 4.40 mm.), base not incised, tip proportionally longer.

For comparison with *E. alpinus* see the account of that species.

In most places where *E. minimus* and *E. amoenus* occur together they can be distinguished without recourse to the baculum, but at Banff and Canmore in western Alberta, recourse to the baculum is almost necessary. There, as elsewhere, they can be distinguished readily by the shape of the bacula.

Specimens examined: 72.

Eutamias minimus borealis: Alberta: Canmore, 1 (BS).

E. m. cacodemus: South Dakota: *Shannon Co.*: Quinn's Draw, Cheyenne River, 1 (NM); 14 mi. N and 5 mi. W Rockyford, 3,200 ft., 1.

E. m. confinis: Wyoming: *Big Horn Co.*: 17 mi. E and 3 mi. S Shell, 9,000 ft., 1; 9 mi. E and 9 mi. N Tensleep, 8,200 ft., 1. *Washakie Co.*: 9 mi. E and 4 mi. N Tensleep, 7,000 ft., 1.

E. m. consobrinus: Montana: *Madison Co.*: 26 mi. NW West Yellowstone, 6,100 ft., 1. Wyoming: *Sublette Co.*: 5 mi. E and 9 mi. N Pinedale, 9,100 ft., 2. *Uinta Co.*: 10 mi. S and 1 mi. W Robertson, 8,700 ft., 1; 13 mi. S and 2 mi. E Robertson, 9,200 ft., 2. Utah: *Uintah Co.*: Paradise Park, 21 mi. W and 15 mi. N Vernal, 10,050 ft., 8. Colorado: *Jackson Co.*: 9-1/2 mi. W and 2 mi. N Walden, 8,400 ft., 1.

E. m. jacksoni: Michigan: *Menominee Co.*: 7 mi. E Stephenson, 4 (MM). Wisconsin: *Juneau Co.*: Camp Douglas, 1 (NM).

E. m. minimus: Wyoming: *Natrona Co.*: 27 mi. N and 1 mi. E Powder River, 6,075 ft., 1; 16 mi. S and 11 mi. W Waltman, 6,950 ft., 1; Sun Ranch, 5 mi. W Independence Rock, 6,000 ft., 1. *Uinta Co.*: 8-1/2 mi. W Ft. Bridger, 7,100 ft., 1; 2 mi. W Ft. Bridger, 6,700 ft., 1; unspecified, 3 (MM). *Sweetwater Co.*: Kinney Ranch, 21 mi. S Bitter Creek, 6,800 ft., 1; 32 mi. S and 22 mi. E Rock Springs, 7,025 ft., 3; 33 mi. S Bitter Creek, 1.

E. m. operarius: Wyoming: *Converse Co.*: 21-1/2 mi. S and 24-1/2 mi. W Douglas, 1. *Carbon Co.*: 8 mi. N and 19-1/2 mi. E Savery, 8,800 ft., 1; 5 mi. N and 5 mi. E Savery, 6,900 ft., 1. *Albany Co.*: 3 mi. ESE Browns Peak, 10,000 ft., 1. Colorado: *Rocky Mountain National Park*, 1 (MM). *Boulder Co.*: Unspecified, 1 (NM). *Gunnison Co.*: 7 mi. S and 7 mi. W Gunnison, 8,150 ft., 1. *Saguache Co.*: 5 mi. N and 22 mi. W Saguache, 10,000 ft., 2. *Archuleta Co.*: Unspecified, 1 (MM); 5 mi. S and 25 mi. W Antonito, 9,600 ft., 3. *Costilla Co.*: Unspecified, 1 (MM). New Mexico: *Taos Co.*: 23 mi. S and 6 mi. E Taos, 8,750 ft., 3.

E. m. pallidus: Montana: *Fergus Co.*: Unspecified, 1 (MM). *Sweetgrass Co.*: Unspecified, 1 (MM). Wyoming: *Campbell Co.*: 4 mi. S and 3 mi. W Rockypoint, 1; Ivy Creek, 8 mi. W and 5 mi. N Spotted Horse, 3; Middle Butte, 6,010 ft., 38 mi. S and 19 mi. W Gillette, 1.

E. m. silvaticus: South Dakota: *Pennington Co.*: Unspecified, 1 (MM). *Custer Co.*: Unspecified, 2 (MM). Wyoming: *Weston Co.*: 1-1/2 mi. E Buckhorn, 6,150 ft., 3.

Eutamias townsendii (Bachman)

Figure 3

Pelage tawny to olivaceous; stripes obscure and underparts tawny, or stripes conspicuous and underparts white; tail slender and sparsely haired; size large; skull largest in the subgenus *Neotamias*.

Baculum: Shaft thin; keel low, 1/5 of length of tip; tip 32 per cent of length of shaft; angle formed by tip and shaft 130°; distal 1/5 of shaft slightly compressed laterally; base deeper and wider than shaft; shaft short, 2.24 mm.

Differs from *E. speciosus*, *E. panamintinus*, *E. umbrinus*, *E. palmeri*, and *E. bulleri*, in base not markedly widened, shaft thinner, tip proportionally shorter; from *E. quadrivittatus*, *E. ruficaudus*, *E. cinereicollis*, and *E. quadrimaculatus*, in shaft thinner, baculum shorter and smaller; from *E. sonomae*, in ridges on either side of tip not enlarged, keel proportionally higher; from *E. amoenus* and *E. dorsalis*, in base widened and thickened, baculum usually shorter; from *E. merriami*, in being markedly shorter, and having base widened and deepened but not incised dorsally.

For comparisons with *E. alpinus* and *E. minimus* see the accounts of those species.

Specimens examined: 2.

E. townsendii townsendii: Oregon: *Multnomah Co.*: Portland, 1 (NM).

E. t. cooperi: Oregon: *Hood Co.*: Brooks Meadow, 4,300 ft., 9 mi. ENE Mount Hood, 1.

Eutamias sonomae Grinnell

Figure 4

Upper parts rich reddish, more or less dulled by gray; backs of pinnae of ears nearly bare; tail long and bushy; size large; skull large, long, and narrow; rostrum deep.

Baculum: Shaft thin; keel low, 1/10 of length of tip; tip 27 to 31 per cent of length of shaft; angle formed by tip and shaft 130°; distal 1/4 of shaft slightly compressed laterally; base deeper and wider than shaft; shaft of medium length, 3.03 to 3.30 mm.; ridges on either side of tip strongly developed, partly obscuring keel from side.

Differs from *E. speciosus*, *E. panamintinus*, *E. umbrinus*, *E. palmeri*, and *E. bulleri*, in ridges on either side of tip strongly developed, shaft thin, base not markedly widened, tip proportionally shorter; from *E. quadrivittatus*, *E. ruficaudus*, *E. cinereicollis*, and *E. quadrimaculatus*, in shaft thin, ridges on either side of tip strongly developed, baculum shorter; from *E. amoenus* and *E. dorsalis*, in keel lower, ridges on either side of tip strongly developed, base thicker and wider; from *E. merriami*, in markedly shorter, ridges on either side of tip strongly developed, tip proportionally longer, base wider and deeper but not incised dorsally.

For comparisons with *E. alpinus*, *E. minimus*, and *E. townsendii*, see the accounts of those species.

Specimens examined: 5.

E. sonomae alleni: California: *Marin Co.*: Unspecified, 1 (NM).

E. sonomae sonomae: California: *Siskiyou Co.*: Salmon Mts., W slope, W of Etna, 1 (BS). *Shasta Co.*: Redding, 1 (BS). *Mendocino Co.*: Cahto, 1 (BS).

Eutamias amoenus (J. A. Allen)

Figure 5

Upper parts generally ochraceous; venter frequently buffy; size small to medium; zygomatic arches usually appressed to cranium; frequently difficult to distinguish from *E. minimus*.

Baculum: Shaft thin; keel low, 1/5 of length of tip; tip 30 to 35 per cent of length of shaft; angle formed by tip and shaft 130°; distal 1/5 of shaft slightly compressed laterally; shaft short, 2.37 to 2.96 mm.

Differs from *E. speciosus*, *E. panamintinus*, *E. umbrinus*, *E. palmeri*, and *E. bulleri*, in base not markedly widened, shaft thinner; from *E. quadrivittatus*, *E. ruficaudus*, *E. cinereicollis* and *E. quadrimaculatus*, in shaft thinner, baculum usually shorter, tip shorter; from *E. dorsalis*, in keel lower, angle formed by tip and shaft 130°, usually smaller; from *E. merriami*, in being markedly shorter, and in base not being dorsally incised, tip proportionally longer.

For comparisons with *E. alpinus*, *E. minimus*, *E. townsendii*, and *E. sonomae*, see the accounts of those species.

Bacula from skins labeled as *E. amoenus amoenus* from, Lardo, Valley Co., Idaho (MM), Butte Co., Idaho (MM), and Boise National Forest, Idaho (BS), resemble the bacula of *E. umbrinus*, and critical study of other features of these specimens probably will show that they are *E. umbrinus*.

Specimens examined: 23.

E. amoenus amoenus: Oregon: *Klamath Co.*: Fort Klamath, 1 (MM). Idaho: *Valley Co.*: Lardo, 1 (BS). *Butte Co.*: Unspecified, 1 (MM). Boise National Forest, 1 (BS).

E. a. ludibundus: British Columbia: Moose Lake, 1 (NM).

E. a. luteiventris: Alberta: Banff, 2 (BS). Montana: *Flathead Co.*: 1 mi. W and 2 mi. S Summit, 5,000 ft., 1. *Teton Co.*: 17-1/8 mi. W and 6-1/2 mi. N Augusta, 5,100 ft., 2. *Missoula Co.*: Unspecified, 1 (MM). Idaho: *Idaho Co.*: 20 mi. S and 2 mi. E Grangeville, 2. *Fremont Co.*: 7 mi. W West Yellowstone, 7,000 ft., 5; 9 mi. SW West Yellowstone, 8,500 ft., 1. Wyoming: *Teton Co.*: 2-1/2 mi. N and 3-1/2 mi. E Moran, 7,225 ft., 1.

E. a. monoensis: California: *Mono Co.*: Leevining Creek, 1 (NM).

E. a. vallicola: Montana: *Ravalli Co.*: Bass Creek, 3-1/2 mi. NW Stevensville, 3,750 ft., 2.

Eutamias dorsalis (Baird)

Figure 6

General tone of upper parts smoky gray; dorsal stripes indistinct or obsolete; skull large.

Baculum: Shaft thin; keel low, 1/5 of length of tip; tip 29 to 40 per cent of length of shaft; angle formed by tip and shaft 140°; distal 1/2 of shaft slightly compressed laterally; shaft short to medium, 2.64 to 3.69 mm.

Differs from *E. speciosus*, *E. panamintinus*, *E. umbrinus*, *E. palmeri*, and *E. bulleri*, in base not widened, shaft thinner; from *E. quadrivittatus*, *E. ruficaudus*, *E. cinereicollis*, and *E. quadrimaculatus*, in shaft thinner, baculum usually shorter; from *E. merriami*, in shaft markedly shorter, base not incised dorsally, tip proportionally longer.

For comparisons with *E. alpinus*, *E. minimus*, *E. townsendii*, *E. sonomae*, and *E. amoenus*, see the accounts of those species.

Specimens examined: 12.

E. dorsalis dorsalis: Arizona: *Yavapai Co.*: 3 mi. N Ft. Whipple, 5,000 ft., 1 (BS). *Gila Co.*: Carr's Ranch, Sierra Ancha Mountains, 1 (BS). *Pima Co.*: Unspecified, 1 (MM). New Mexico: *Valencia Co.*: Mount Taylor, 1 (MM); 1 mi. N Cebolleta, 1 (MM). *Socorro Co.*: San Mateo Mountains, 1 (BS). *Chihuahua*: Sierra Madre, near Guadalupe y Calvo, 3 (BS).

E. d. utahensis: Wyoming: *Sweetwater Co.*: W side Green River, 1 mi. N Utah Border, 1. Colorado: *Moffat Co.*: Escalante Hills, 20 mi. SE Ladore, 1 (BS). Arizona: *Coconino Co.*: Ryan, 1 (BS).

Eutamias merriami (J. A. Allen)

Figure 7

General tone of upper parts grayish; cheeks and underparts white, more or less dulled by gray; size large; skull large.

Baculum: Shaft thin; keel low, 2/5 of length of tip; tip 16 per cent of length of shaft; angle formed by tip and shaft 130°; distal 1/10 of shaft slightly compressed laterally; base incised dorsally; shaft markedly long, 4.88 mm.

Differs from *E. speciosus*, *E. panamintinus*, *E. umbrinus*, *E. palmeri*, and *E. bulleri*, in base incised dorsally, base not widened, shaft thinner, tip proportionally shorter; from *E. quadrivittatus*, *E. ruficaudus*, *E. cinereicollis*, and *E. quadrimaculatus*, in base incised dorsally, shaft thinner, tip proportionally much shorter.

For comparisons with *E. alpinus*, *E. minimus*, *E. townsendii*, *E. sonomae*, *E. amoenus*, and *E. dorsalis*, see the accounts of those species.

Specimens examined: One from Mount Piños, Ventura Co., California (LA).

Eutamias quadrivittatus (Say)

Figures 8-10

Color bright and tawny; size medium to large; braincase widened.

Baculum of *E. q. quadrivittatus*: Shaft thick; keel proportionally low, 1/4 of length of tip; tip 30 to 44 per cent of length of shaft; angle formed by tip and shaft 130°; distal 1/3 of shaft slightly compressed laterally; shaft long, 3.17 to 3.62 mm.

Differs from *E. q. hopiensis*, in baculum larger, angle formed by tip and shaft less distinct; from *E. ruficaudus ruficaudus*, in shaft and tip proportionally shorter; from *E. r. simulans*, in keel proportionally lower, shaft usually shorter and narrower; from *E. cinereicollis*, in shaft shorter, angle formed by tip and shaft larger, ridges on either side of tip more distinct; from *E. quadrimaculatus*, in tip proportionally shorter, shaft shorter; from *E. speciosus*, *E. panamintinus*, *E. umbrinus*, *E. palmeri*, and *E. bulleri*, in base not markedly widened, shaft longer, angle formed by tip and shaft larger.

For comparisons with *E. alpinus*, *E. minimus*, *E. townsendii*, *E. sonomae*, *E. amoenus*, *E. dorsalis*, and *E. merriami*, see the accounts of those species.

Baculum of *E. q. hopiensis*: Shaft thick; keel proportionally low, 1/3 of length of tip; tip 34 per cent of length of shaft; angle formed by tip and shaft 130°; distal 1/3 of shaft slightly compressed laterally; shaft short, 2.64 mm. in length.

Differs from *E. ruficaudus ruficaudus*, *E. r. simulans*, *E. cinereicollis*, and *E. quadrimaculatus*, in being shorter, proportions as in *E. q. quadrivittatus*; from *E. speciosus*, *E. panamintinus*, *E. umbrinus*, *E. palmeri*, and *E. bulleri*, in base not markedly widened, angle formed by tip and shaft larger; from *E. alpinus*, *E. minimus*, *E. townsendii*, *E. sonomae*, *E. amoenus*, *E. dorsalis*, and *E. merriami*, except for smaller size, as in *E. q. quadrivittatus*.

Bacula of *E. q. hopiensis* from northeastern Arizona are like those in typical *E. q. quadrivittatus*. The specimens from this region, to judge from parts of the animal other than the baculum, are intergrades between *E. q. hopiensis* and *E. q. quadrivittatus*. Specimens from near Moab, Grand Co., Utah, are typical *E. q. hopiensis* and the bacula of these specimens are considerably smaller than those of specimens of typical *E. q. quadrivittatus*. No bacula are known to me that are structurally intermediate between those of *E. q. quadrivittatus* and *E. q. hopiensis*.

Specimens examined: 21.

E. quadrivittatus quadrivittatus: Colorado: *Gunnison Co.*: Sapinero, 1 (BS). *Chaffee Co.*: Unspecified, 1 (MM). *Saguache Co.*: 5 mi. N and 22 mi. W Saguache, 10,000 ft., 1. *Fremont Co.*: Canon City, 3 (BS). New Mexico: *Rio Arriba Co.*: Dulce, 1 (BS). *Taos Co.*: 23 mi. S and 6 mi. E Taos, 8,750 ft., 4. *Union Co.*: Sierra Grande, 1 (BS). *Valencia Co.*: Mirror Spring, Mt. Taylor, 3 (MM). *Torrance Co.*: S end, E slope, Manzano Mountains, 2 (BS).

E. quadrivittatus hopiensis: Utah: *Grand Co.*: Mouth of Nigger Bill Canyon, E side Colorado River, 4 mi. N Moab Bridge, 4,500 ft., 1 (UU); Moab, 4,500 ft., 1 (UU). Arizona: *Apache Co.*: Tunicha Mountains, 1 (BS).

Eutamias ruficaudus Howell

Figures 11-12

General tone of upper parts dark tawny; size medium; braincase narrowed.

Baculum of *E. r. ruficaudus*: Shaft thick; keel proportionally low, 1/4 of length of tip; tip 31 to 33 per cent of length of shaft; angle formed by tip and shaft 120°; distal 2/5 of shaft slightly compressed laterally; base slightly widened; shaft long, 4.09 to 4.56 mm.

Differs from *E. r. simulans*, in keel proportionally lower, tip proportionally shorter, distal 1/5 of shaft less laterally compressed, shaft usually longer; from *E. cinereicollis*, in angle formed by tip and shaft smaller, shaft shorter, ridges on either side of tip distinct; from *E. quadrimaculatus*, in tip proportionally longer, shaft usually shorter; from *E. speciosus*, *E. panamintinus*, *E. umbrinus*, *E. palmeri*, and *E. bulleri*, in base not markedly widened, shaft longer, angle formed by tip and shaft larger.

For comparisons with *E. alpinus*, *E. minimus*, *E. townsendii*, *E. sonomae*, *E. amoenus*, *E. dorsalis*, *E. merriami*, and *E. quadrivittatus*, see the accounts of those species.

Baculum of *E. r. simulans*: Shaft thick; keel proportionally low, 2/5 of length of tip; tip 40 to 48 per cent of length of shaft; angle formed by tip and shaft

130°; distal 2/5 of shaft laterally compressed; base slightly wider than shaft; shaft medium to long, 3.30 to 4.26 mm.

Differs from *E. cinereicollis* in, keel proportionally higher, shaft shorter, tip proportionally longer, angle formed by tip and shaft smaller, distal 2/5 of shaft more laterally compressed; from *E. quadrimaculatus* in, keel proportionally higher, angle formed by tip and shaft larger, tip proportionally longer, distal 2/5 of shaft more laterally compressed, shaft shorter; from *E. speciosus*, *E. panamintinus*, *E. umbrinus*, *E. palmeri*, and *E. bulleri*, in base not markedly widened, shaft usually longer.

For comparisons with *E. alpinus*, *E. minimus*, *E. townsendii*, *E. sonomae*, *E. amoenus*, *E. dorsalis*, *E. merriami*, and *E. quadrivittatus*, see the accounts of those species.

The differences between the bacula of the subspecies *Eutamias ruficaudus ruficaudus* and *E. r. simulans* are comparable to those usually found between species of chipmunks. Consequently, I suspect that *E. r. simulans* and *E. r. ruficaudus* are specifically distinct and suggest that a search would be worth while for specimens in the geographic area between the geographic ranges as now known for the two kinds to ascertain whether intergradation (the criterion of subspecies) occurs. I suppose there is no intergradation but in the absence of precise information, I choose not to modify the current taxonomic arrangement of *E. r. ruficaudus* and *E. r. simulans*.

Specimens examined: 17.

E. ruficaudus ruficaudus: Montana: *Flathead Co.*: 1 mi. W and 2 mi. S Summit, 5,000 ft., 2. *Ravalli Co.*: Big Hole Hill, 6,000 ft., 2; Big Hole Hill, 6,600 ft., 1; Continental Divide, Big Hole Hill, 7,000 ft., 1.

E. r. simulans: Idaho: *Bonner Co.*: Priest Lake, 1 (BS). *Kootenai Co.*: 13 mi. E and 5 mi. N Coeur d'Alene, 3. *Shoshone Co.*: Mullan, 1 (BS). *Clearwater Co.*: 25 mi. E and 16 mi. N Pierce, 6.

Eutamias cinereicollis (J. A. Allen)

Figure 13

General tone of upper parts dark grayish tawny; size medium to large; skull large; braincase widened.

Baculum: Shaft thick; keel proportionally low, 1/5 of length of tip; tip 34 per cent of length of shaft; angle formed by tip and shaft 145°; distal 1/3 of shaft slightly compressed laterally; base slightly widened; shaft long, 4.88 mm.

Differs from *E. quadrimaculatus*, in keel lower, ridges on either side of tip weakly developed, angle formed by tip and shaft larger; from *E. speciosus*, *E. panamintinus*, *E. umbrinus*, *E. palmeri*, and *E. bulleri*, in shaft longer, base not markedly widened, angle formed by tip and shaft much larger.

For comparisons with *E. alpinus*, *E. minimus*, *E. townsendii*, *E. sonomae*, *E. amoenus*, *E. dorsalis*, *E. merriami*, *E. quadrivittatus*, and *E. ruficaudus*, see the accounts of those species.

Specimen examined: One from Mount Thomas, White Mountains, Apache Co., Arizona (BS).

Eutamias quadrimaculatus (Gray)

Figure 14

General tone of upper parts bright reddish; pattern inconspicuous; light and dark facial stripes strongly contrasting; size large; skull relatively small and slightly built.

Baculum: Shaft thick; keel relatively low, 1/4 of length of tip; tip 27 per cent of length of shaft; angle formed by tip and shaft 120°; distal 1/3 of shaft slightly compressed laterally; base slightly wider than shaft; shaft long, 4.35 to 5.28 mm.

Differs from *E. speciosus*, *E. panamintinus*, *E. townsendii*, *E. umbrinus*, *E. palmeri*, and *E. bulleri*, in shaft markedly longer; base not markedly widened; angle formed by tip and shaft larger; tip proportionally shorter.

For comparisons with all other species of chipmunks from western North America, see the accounts of those species.

Specimens examined: 4.

California: *Plumas Co.*: Mountains near Quincy, 1 (BS). *Placer Co.*: Blue Canyon, 1 (NM). *Alpine Co.*: Markleeville, 1 (BS); N fork Stanislaus River, 1 (BS).

Eutamias speciosus (Merriam)

Figure 15

General tone of upper parts bright; light and dark elements of color pattern strongly contrasting; outer stripes broad and strikingly conspicuous; size medium; skull moderately broadened; dorsal outline of skull strongly arched in profile; upper incisors short and sharply recurved.

Baculum: Shaft thick; keel of medium height, 1/3 of length of tip; base of keel 1/3 of length of tip; tip 47 to 55 per cent of length of shaft; angle formed by tip and shaft 90°; distal 2/3 of shaft markedly compressed laterally; base markedly wider than shaft; shaft short to medium, 2.11 to 3.17 mm.

Differs from *E. panamintinus*, in base of keel proportionally shorter, angle formed by tip and shaft smaller; from *E. umbrinus* and *E. palmeri*, in keel higher, angle formed by tip and shaft smaller, distal 2/3 (not 1/2) of shaft markedly laterally compressed, base markedly wider; from *E. bulleri*, in keel smaller, shaft shorter, tip proportionally longer, ridges on either side of tip distinct.

For comparisons with all other species of chipmunks of western North America, see the accounts of those species.

The baculum in *E. speciosus frater* is approximately the same size as in *E. umbrinus inyoensis*, but differs in shape as described above.

Specimens examined: 6.

E. speciosus frater: California: *Madera Co.*: San Joaquin River, near head of N fork, Sierra Nevada Mountains, 2 (BS).

E. speciosus sequoiensis: California: *Tulare Co.*: Mount Whitney, 1 (BS); Sequoia National Park, 2 (BS).

E. speciosus callipeplus: California: *Kern Co.*: Mount Pinos, 1 (BS).

Eutamias panamintinus (Merriam)

Figure 16

Dorsal dark stripes reddish; size small to medium; skull of medium size; braincase flattened.

Baculum: Shaft thick; keel low, 1/3 of length of tip; base of keel 1/2 of length of tip; tip 52 per cent of length of shaft; angle formed by tip and shaft 110°; distal 2/3 of shaft moderately compressed laterally; base markedly widened; shaft short, 2.17 mm.

Differs from *E. umbrinus* and *E. palmeri*, in distal 2/3 (rather than 1/2) of shaft moderately compressed laterally; from *E. bulleri*, in keel smaller, shaft shorter, ridges on either side of tip distinct.

For comparisons with other species of chipmunks of western North America, see the accounts of those species.

The structure of the baculum most closely resembles that of *E. speciosus* and the geographic ranges of these two species are in juxtaposition.

Specimen examined: One from Coal Kilns, Panamint Mountains, Inyo Co., California (CN).

Eutamias umbrinus (J. A. Allen)

Figures 17-18

General tone of upper parts dark; size medium to large; skull of medium size; braincase narrowed.

Baculum: Shaft thick; keel low, 1/4 of length of tip; tip 36 to 50 per cent of length of shaft; angle formed by tip and shaft 100°; distal 1/2 of shaft markedly compressed laterally; base markedly widened; shaft short to medium, 2.51 to 3.03 mm.

Differs from *E. bulleri*, in shaft shorter, keel smaller, ridges on either side of tip distinct; distal 1/2 of shaft strongly compressed laterally. Does not differ from *E. palmeri*.

For comparisons with all other species of chipmunks of western North America, see the accounts of those species.

Specimens of *E. umbrinus montanus* from north-central Colorado have, in the past (Howell 1929:82), been referred to *E. quadrivittatus quadrivittatus*. In many features these two kinds of chipmunks resemble each other closely; their bacula, nevertheless, differ markedly (compare figs. 8-9 with 17-18).

Specimens examined: 25.

E. umbrinus umbrinus: Wyoming: *Uinta Co.*: 10 mi. S and 1 mi. W Robertson, 8,700 ft., 1. Utah: *Uintah Co.*: Paradise Park, 21 mi. W and 15 mi. N Vernal, 10,050 ft., 12.

E. u. adsitus: Utah: *Beaver Co.*: Britts Meadow, Beaver Range Mountains, 8,500 ft., 1 (BS). *Wayne Co.*: Donkey Lake, Boulder Mountain, 10,000 ft., 1 (UU). *Garfield Co.*: Wildcat R. S., Boulder Mountain, 8,700 ft., 1 (UU).

E. u. sedulus: Utah: *Garfield Co.*: Mount Ellen, Henry Mountains, 1 (BS).

E. u. inyoensis: California: *Tulare Co.*: Mount Whitney, head of Big Cottonwood Creek, 2 (BS). Nevada: *Elko Co.*: W side Ruby Lake, 3 mi. N Elko County Line, 1.

E. u. nevadensis: Nevada: *Clark Co.*: Sheep Mountains, 1 (MM).

E. u. fremonti: Wyoming: *Sublette Co.*: 31 mi. N Pinedale, 8,025 ft., 1; 19 mi. W and 2 mi. S Big Piney, 1.

E. u. montanus: Colorado: *Boulder Co.*: 3 mi. S Ward, 9,000 ft., 1; 1/2 mi. E and 3 mi. S Ward, 9,400 ft., 1.

Eutamias palmeri Merriam

Figures 17-18

General tone of upper parts grayish tawny; ocular stripe pale; skull, rostrum, nasals, and upper incisors shorter than in *E. umbrinus*.

Baculum: Indistinguishable from that of *E. umbrinus*. This supports the opinion of previous students that *E. palmeri* is a close relative of *E. umbrinus* which occurs immediately to the north and east. Intergradation does not occur between these two species, for, low-lying terrain, inhospitable to chipmunks, isolates *E. palmeri* from its relatives. (Verbal information from E. R. Hall.)

Specimen examined: One from Charleston Peak, 8,000 ft., Clear Creek Co., Nevada.

Eutamias bulleri (J. A. Allen)

Figure 19

General tone of upper parts dark; dorsal dark stripes conspicuous and black; size large; skull large.

Baculum: Shaft thick; keel high, 1/2 of length of tip; keel long, 1/2 of length of tip; tip 40 to 48 per cent of length of shaft; angle formed by tip and shaft 100°; base markedly widened; shaft of medium length, 3.30 mm.

For comparisons with all other species of chipmunks of western North America, see the accounts of those species.

The large size of the keel of the baculum in this species is distinctive among chipmunks of western North America.

Specimens examined: 2.

E. bulleri bulleri: Zacatecas: Sierra de Valparaiso, 2 (NM).

FIGS. 1-19. Lateral view of right side, unless otherwise indicated, of the baculum in each of the species of chipmunks (subgenus *Neotamias*) of western North America:

1. *Eutamias alpinus*, No. 12577 CN; from Big Cottonwood Meadow, Tulare Co., California.

2. *E. minimus consobrinus*, No. 25439; from 13 mi. S and 2 mi. E Robertson, 9,200 ft., Uinta Co., Wyoming.

3. *E. townsendii cooperi*, No. 53169; from Brooks Meadow, 4,300 ft., 9 mi. ENE Mt. Hood, Hood River Co., Oregon.

4. *E. sonomae sonomae*, No. 98436 BS; from Redding, Shasta Co., California.

5. *E. amoenus luteiventris*, No. 33811; from 7 mi. W West Yellowstone, 7,000 ft., Fremont Co., Idaho.

6. *E. dorsalis dorsalis*, No. 213415 BS; from 3 mi. N Ft. Whipple, 5,000 ft., Yavapai Co., Arizona.

7. *E. merriami merriami*, No. 1270 LA; from Mount Pinos, Ventura Co., California.

8 and 9. *E. quadrivittatus quadrivittatus*, No. 35648/47919 BS; from Canon City, Fremont Co., Colorado. Figure 9 in dorsal view.

10. *E. quadrivittatus hopiensis*, No. 783 UU; from Moab, 4,500 ft., Grand Co., Utah.

11. *E. ruficaudus ruficaudus*, No. 33884; from 1 mi. W and 2 mi. S Summit, 5,000 ft., Flathead Co., Montana.

12. *E. ruficaudus simulans*, No. 41478; from 13 mi. E and 5 mi. N Coeur d'Alene, Kootenai Co., Idaho.

13. *E. cinereicollis cinereicollis*, No. 208621 BS; from Mount Thomas, White Mountains, Apache Co., Arizona.

14. *E. quadrimaculatus*, No. 95780 BS; from Mountains near Quincy, Plumas Co., California.

15. *E. speciosus sequoiensis*, No. 29135/41203 BS; from Mount Whitney, Tulare Co., California.

16. *E. panamintinus panamintinus*, No. 12502 CN; from Coal Kilns, Panamint Mountains, Inyo Co., California.

17. *E. umbrinus umbrinus*, No. 38062; from Paradise Park, 21 mi. W and 15 mi. N Vernal, 10,050 ft., Uintah Co., Utah.

18. *E. umbrinus montanus*, No. 20105; from 1/2 mi. E and 3 mi. S Ward, 9,400 ft., Boulder Co., Colorado. Dorsal view.

19. *E. bulleri bulleri*, No. 193142 NM; from Sierra del Valparaiso, Zacatecas.

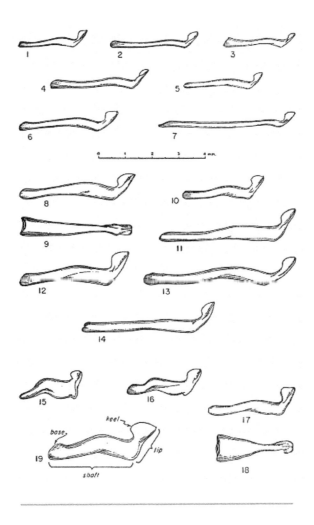

Discussion

In California, Johnson (1943) recognized ten species of chipmunks and assigned these to the five main groups of species which were proposed by Howell (1929). In characterizing each species, Johnson (*op. cit.*) not only made a careful study of skins and skulls, but also employed many ecological data.

Study of the bacula of the Californian chipmunks supports Johnson's (*op. cit.*) conclusion that there are ten species, but suggests that there are three (not five) groups of species in California—as well as elsewhere within the geographic range of the subgenus *Neotamias*. The three groups are (see figs. 1-19): 1. *minimus*-group (*E. alpinus*, *E. minimus*, *E. townsendii*, *E. sonomae*, *E.*

amoenus, *E. dorsalis*, and *E. merriami*); 2. *quadrivittatus*-group (*E. quadrivittatus*, *E. ruficaudus*, *E. cinereicollis*, and *E. quadrimaculatus*); and 3. *speciosus*-group (*E. speciosus*, *E. panamintinus*, *E. umbrinus*, *E. palmeri*, and *E. bulleri*).

Eutamias panamintinus, according to Howell (*op. cit.*:78) and Johnson (*op. cit.*:83), is a near relative of *E. amoenus*. But, the baculum in *E. panamintinus* more closely resembles that in *E. speciosus* than that in *E. amoenus* (compare figs. 5, 15, and 16). Consequently I have placed *E. panamintinus* in the *speciosus*-group.

In north-central Colorado, specimens that really are *E. umbrinus* (subspecies *montanus*) have, in the past (Howell *op. cit.*:82), been referred to *E. quadrivittatus quadrivittatus*, but the bacula of the two species differ markedly from each other (compare figs. 8-9 with 17-18) and permit the specimens readily to be correctly identified to species. Further, Howell (*op. cit.*:95) placed *E. umbrinus* (subspecies *umbrinus* and *fremonti* of current usage) in the *quadrivittatus*-group, whereas the structure of the baculum leads me to place *E. umbrinus* in the *speciosus*-group.

Thus, groups of species established on the basis of only skulls and skins, in a few instances differ from those established on a broader basis which includes the bacula.

Johnson (*op. cit.*:63) writes, "Each species [of Eutamias] has a characteristic habitat which differs from those of other species. Where two or more species occur together in a general locality they are usually mutually exclusive in their choice of foraging and nesting sites and in the time of breeding." Thus he classified the species of Californian chipmunks not only by morphologic characteristics but by habits and habitats as well. The characteristics of the skulls and skins of chipmunks probably reflect the habitats in which these animals live. The characteristics of the bacula, on the other hand, may also reflect the habitats in which the animals live, but to a lesser degree. Because the structures of the bacula are probably less affected by the action of the external environment they probably indicate relationships between groups of species of chipmunks more clearly than do characteristics of the skulls and skins.

If the structures of the bacula indicate relationships between groups of species of chipmunks more clearly than do the characteristics of the skulls and skins, the close resemblance of the skulls of *E. quadrivittatus* and *E. umbrinus* may be thought of as convergence. The same can be said of *E. amoenus* and *E. panamintinus*.

LITERATURE CITED

FRILEY, C. E., JR.

1947. Preparation and preservation of the baculum of mammals. Jour. Mamm., 28:395-397, 1 fig., December 1.

HOWELL, A. H.

1929. Revision of the American chipmunks (genera *Tamias* and *Eutamias*). N. Amer. Fauna, 52:1-157, 10 pls., 9 maps.

JOHNSON, D. H.

1943. Systematic review of the chipmunks (genus Eutamias of California). Univ. California Publ. Zool., 48:63-148, 6 pls., 12 figs, in text, December 24.

WHITE, J. A.

1951. A practical method for mounting the bacula of small mammals. Jour. Mamm., 32:125, February 15.

Transmitted June 26, 1953.

24-8968

www.ingramcontent.com/pod-product-compliance
Ingram Content Group UK Ltd.
Pitfield, Milton Keynes, MK11 3LW, UK
UKHW031340260325
456749UK00002B/289